Water o

DENYS GAMBLIN

cartoons by Field

THOMAS TELFORD LTD
LONDON 1979

Published by Thomas Telford Ltd, Telford House, PO Box 101,
26–34 Old Street, London EC1P 1JH
© Thomas Telford Ltd and Denys Gamblin, 1979

ISBN: 0 7277 0070 7

FOREWORD

It is strange to consider that the origin of this book was in a brief conversation with two individuals who were currently involved in the genesis of the magazine which became *New Civil Engineer*. In a weak moment I agreed that I could probably produce about twenty articles in the Waterboard Files format for this new brainchild.

In the event, I managed to produce over sixty episodes—which received a somewhat mixed reception. In a readership poll the series managed to come out top in both the Most Popular and Least Popular classes, which must be unique. I hope the selection provided here will give pleasure to those who enjoyed them, a twinge of nostalgic pain to those who didn't, and amusement to those who have become readers of *New Civil Engineer* in recent years.

Since *NCE* began in 1972 there have been many changes in the water industry—and the discerning reader may pinpoint the dividing line following the 1974 reorganization, when I stopped guiding the fortunes of a water board and assumed the apprehensive cloak of a divisional manager. One thing which does not change much is human nature and the engineering problems that result from its vagaries. However, to all my friends I am happy to confirm that only my villains are fictitious.

Denys Gamblin

Miss Cunningham was worried about feathers in the water

1

Water Engineer and Manager
Water Board Offices

Dear Sir,
There is something wrong with your water supply
because I keep getting feathers through the tap when I want a
glass of water.
Please look into this complaint immediately because I am
getting very worried.

Yours faithfully,
Agatha Cunningham (Miss)

INTERNAL MEMORANDUM
5 May

To: Distribution Engineer   From: Engineer and Manager
I enclose a copy of a letter dated 4 May from Miss Agatha
Cunningham. Is someone pulling our legs about this? In any
case we shall have to investigate. Instruct the local water
inspector to call on some pretext – checking on pressures
perhaps – and send me his report.

INTERNAL MEMORANDUM
7 May

To: Engineer and Manager   From: Distribution Engineer

Fairhurst, Panton Gardens
Inspector Ffaulkes called at the above address today and his
report is attached: 'The old lady was out but the maid who let me

in said that Miss Cunningham was worried about feathers in the water. Seemed all right to me – clear, 90 ft head pressure, comes directly off main from filters on inlet to Panton service reservoir. Miss Cunningham will be in on Monday and Tuesday (10 and 11) if you want to speak to her.

Water Board Offices
10 May

Miss Agatha Cunningham
Fairhurst
Panton Gardens

Dear Madam,

   Thank you for your letter of 4 May complaining about feathers in your water supply. I'm sure you must be mistaken about this because your supply is straight off the main from the filters and nothing could get into the water. Perhaps there was a feather in the glass before you filled it with water?

  Nevertheless, I will ask the Waterworks Chemist to call at your house at 3 pm on Tuesday, 11 May to take a water sample and reassure you.

Yours faithfully,
Water Engineer and Manager

INTERNAL MEMORANDUM
10 May

To: Chemist and Bacteriologist From: Engineer and Manager
Please call at Fairhurst, Panton Gardens at 3 pm on Tuesday, 11 May to see Miss Cunningham who complains of feathers in her water supply.

  Better arrange for a C and B sample at the same time.

INTERNAL MEMORANDUM
14 May

To: Engineer and Manager From: Chemist and Bacteriologist
Fairhurst, Panton Gardens
I saw Miss Cunningham (an intelligent lady) on Tuesday and

2

explained why it is impossible for feathers to appear in her water supply. The tap was left running for 10 minutes while I was there with no result. Samples were taken and analyses are enclosed – quite typical for this source apart from pH which is too low at 7.2

Fairhurst
Panton Gardens
17 May

Water Engineer and Manager
Water Board Offices.

Dear Sir,

It was kind of you to send the Chemist to my house – such a nice sincere young man – but I'm afraid that he does not appreciate the problem.

Another feather came out of the tap last night and I am sending it down to you with this letter. I have asked my maid to give it to you personally.

I'm sure you will know what to do.

Yours faithfully,
Agatha Cunningham (Miss)

INTERNAL MEMORANDUM
18 May

To: Distribution Engineer   From: Engineer and Manager
(copy to Chemist)

Fairhurst, Panton Gardens

I don't know why I have to do the jobs of you chaps as well as my own and really I should not be scrambling about in this fashion at my age. However – I called on Miss Cunningham on my way to the office this morning and it turns out that she drinks a glass of water before she goes to bed. She draws this water from the basin in the bathroom (a practice she won't repeat in future) and this tap is fed by the tank in the false roof. As is often the case with old properties, it is a galvanized tank without a cover and I discovered a dead owl in the tank. The owl has been removed but please arrange to install a new covered tank at this address; consumer's expense.

All bodies discovered in Deepwet are to be towed downstream with a boat-hook

2

County Police HQ
1 June

Engineer and Manager,
Water Board Offices

Dear Sir,

Deepwet Reservoir

I must protest regarding the false information passed to the men of my Force on Saturday by your reservoir keeper at Deepwet.

He reported a dead body in the reservoir but when the patrol car called at the site it was found that the alleged corpse was within the Willpipe City Boundary and therefore a matter for the City Police Force.

I should be grateful if you would ensure that your employees are aware of the County Boundary to avoid any future waste of time or friction.

Yours faithfully,
Assistant Chief Constable

City Police HQ
1 June

Engineer and Manager
Water Board Offices

Dear Sir,

Deepwet Reservoir

I should be pleased if you would instruct your employees to pass correct information to this office at the earliest opportunity.

A corpse was recovered from the above reservoir on Saturday but only after time had been lost in several telephone conversations with your keeper who did not understand the topography of his reservoir and could not pinpoint the position of the body.

5

Presumably he will be available for the inquest if he can find his way to the City Centre?

Yours faithfully,
Chief Constable

INTERNAL MEMORANDUM
2 June
To: Supplies Engineer   From: Engineer and Manager
Deepwet Reservoir

I attach copies of two letters from our local police forces in connection with the body recovered from Deepwet on Saturday. What lies behind these complaints?

We have had eight suicides in this reservoir during the past six years and I would have thought that the reservoir keeper would have been well aware of the procedure by now!

No address
Water Board Engineer   *Received 3 June*

Dear Sir,

Why did your reservoir keeper take the City Police through the gate into Murrain Wood alongside Deepwet Reservoir on Saturday? This is outside the City area and the County Police are well able to take care of anything inside the County. No wonder our rates are so high with the City Police poking their noses into things that don't concern them and your reservoir keeper up to no good as well and being paid overtime into the bargain.

Yours faithfully,
A Friend

INTERNAL MEMORANDUM
3 June
To: Engineer and Manager   From: Supplies Engineer
Deepwet Reservoir

Rainchart, the reservoir keeper, started his holidays last Friday and his deputy,Fishticket, although a good man, has not dealt with a suicide before. As I understand it, the timetable of events was as follows:

5.30 pm Fishticket notices body in water at Murrain Wood and telephones City Police. City Police say this is inside County area and tell Fishticket to phone County Police.

5.40 pm Fishticket phones County Police who send out a patrol car.

5.55 pm Fishticket takes patrol car to reservoir at Murrain Wood but current of water from Murrain Beck has now carried body out of County area and into City area. County Police annoyed and tell Fishticket to ring City Police.

6.10 pm Fishticket rings City Police (twice – they didn't believe him the first time!) and finally persuades them that corpse is within their jurisdiction.

6.35 pm City Police squad car and ambulance arrive at Deepwet. The body is in the City area but, to recover it, the police had to obtain access to the reservoir via the gate in Murrain Wood within the County area.

I have discreetly informed Fishticket that, in accordance with custom, all bodies in Deepwet are to be towed downstream with a boathook and 'discovered' in the pool beside the overflow.

Water Board Offices
5 June
Chief Constable
Willpipe City Police
(Copy to County Police HQ)

Dear Sir,
                    Deepwet Reservoir
    I must apologise for the time wasted on Saturday with regard to the body recovered from the above reservoir.
    Presumably the corpse was a stranger to the locality!
                    Yours faithfully,
                    Engineer and Manager

**The general public would be invited to taste samples of water from
all the Board's resources**

3

Cutting from the *'Willpipe Express'* dated 19 June.

'During his report to the Board, the Water Engineer and Manager commented on the good progress made in reconditioning Claypack service reservoir. This reservoir, holding 750 000 gallons, was originally constructed in 1908 as an open tank with mass concrete walls. Reconditioning works, including a precast concrete roof and a rubber lining, had begun in March and would be completed by the end of the present month when the reservoir would be returned to service'.

14 Garden Crescent
Claypack
10 July

Water Engineer and Manager
Water Board Offices

Dear Sir,

Claypack Reservoir

I have been instructed by my Association to protest in the strongest possible terms about the Board's reprehensible action in lining Claypack reservoir with rubber. Since the reservoir returned to service, there has been a terrible taste in the water and it is impossible to drink. Many children in the area are now ill and the matter has been brought to the attention of the Medical Officer of Health.

All consumers are now obtaining their drinking water from the local spring or else transporting it from other areas. I require your assurance that action will be taken immediately!

Yours faithfully,
R.T. Champion,
Hon. Secretary
Willpipe Ratepayers Association

To: Engineer and Manager   From: Clerk/Treasurer

Claypack Reservoir

I have received 15 letters and 23 verbal complaints of the rubbery taste of the water from Claypack reservoir. I know that both you and the Chemist have assured the Board that there is no taste and that the water is unaffected by the lining but the public do not agree. What can we do?

Lord Mayor's Parlour
City of Willpipe
14 July

Robert Fitzjohn Esq
Engineer and Manager
Water Board Offices

Dear Mr Fitzjohn,

I have been receiving innumerable complaints about the taste of the water from Claypack reservoir in my capacity as Chairman of the Water Board and I look forward to a comprehensive report from you to the members at the next Board meeting on Wednesday.

The press will be present as usual and public anxiety must be allayed.

Yours faithfully,
L.P. Chainman,
Lord Mayor

Cutting from the *'Willpipe Express'* dated 17 July

'The Water Engineer and Manager commented on the complaints received in connection with the taste of water derived from Claypack service reservoir. He stated that comprehensive analyses of water samples taken both by the Board's Chemist and the Public Health laboratories of the Medical Officer were completely satisfactory and had poss-

essed neither tastes nor odours. Nevertheless, to demonstrate this fact, he would arrange for a testing station to be set up in Claypack. The general public would be invited to taste samples of water from all the Board's sources and to state which sample had been taken from Claypack reservoir'.

INTERNAL MEMORANDUM
17 July
To: Distribution Engineer   From: Engineer and Manager
(Copy to: Waterworks Chemist)

Claypack Tests
Please arrange for the Board's caravan to be set up in the Square at Claypack on Friday and Saturday, 25 and 26 July together with a large notice inviting consumers to taste the water between the hours of 10 am and 9 pm. Inside the caravan, have 4 belljars (labelled A, B, C and D) of water from the Deepwet source taken from the outlets of Panton, Walton, Crawley and Claypack service reservoirs.

A representative from the Medical Officer of Health's department will assist you and a journalist from the Willpipe Express will be present to see fair play. Please rotate the labels on the second day to avoid collusion and arrange to take the names and addresses of all participants.

Water Board Offices
R.T. Champion Esq   28 July
Hon. Sec. Ratepayers Association
14 Garden Crescent
Claypack

Dear Sir,
Claypack Reservoir
You will be aware of the tests carried out in Claypack Square last Friday and Saturday when all consumers were invited to detect the Claypack reservoir water from other samples in the presence of impartial observers. I can now give you the

11

results of the tests which will be published in the Willpipe Express tomorrow. They are:

No. of participants		1533
Claypack water attributed to:	Panton reservoir:	383
	Walton reservoir:	359
	Crawley reservoir:	316
	Claypack reservoir:	317
	Don't know:	158

You will appreciate that, if your allegation had been correct, the number opposite Claypack reservoir should have been greater than all the others and, to my mind, the results demonstrate conclusively that the Claypack water is no different from all other supplies.

Yours faithfully,
Engineer and Manager.

14 Garden Crescent
Claypack
30 July

Engineer and Manager
Water Board Offices

Dear Sir,

Claypack Reservoir
I have your letter of the 28 July and must say that your statistics do not impress me. To my mind(!) the results demonstrate conclusively that all the water supplied by the Water Board tastes as if it came out of a hot water bottle.

Yours faithfully,
R.T. Champion

4

Civil Engineer
British Rail

Dear Sir,

Two Arches Rail Crossing

I write in connection with the 24 in. dia. steel trunk main to be laid under the railway bridge at Two Arches and the conditions you have laid down for its installation. I confirm that the main will have 4 ft of cover and that the trench will be close-boarded and refilled with lean concrete on completion of mainlaying.

However I do object to your condition that the trench must be strutted at 3 ft intervals both vertically and horizontally and that the struts must not be removed while the trench is open. This is completely unrealistic and I should be grateful if you would inform me how 24 in. dia. steel pipes, about 25 ft long, can be placed in the trench under these conditions. As you know, you refused to permit our alternative method of laying by thrust-bore.

Yours faithfully,
Engineer and Manager

Civil Engineer's Office
Willpipe Central Station
16 August

Engineer and Manager
Water Board Offices

Dear Sir,

Two Arches Rail Crossing
I acknowledge your letter of the 14 August in connection with the above but I regret that British Rail cannot relax their

Please inform me how the pipes were installed in the trench

conditions with regard to the laying of the 24 in. main under the railway bridge.

I appreciate your difficulties and, in answer to your query, may I remind you that this is your problem and you must solve it yourself.

Mr Jenkins, my Inspector of Works will be present on site when mainlaying is proceeding in the vicinity of Two Arches to ensure that the conditions are observed.

Yours faithfully,
S.T. Standfast,
Divisional Civil Engineer

INTERNAL MEMORANDUM
18 August
To: New Works Engineer   From: Engineer and Manager

Two Arches Rail Crossing

I attach copies of correspondence with British Rail in connection with the above. You will note that they refuse to relax their conditions and, in addition, you will now be saddled with a watchdog named Jenkins.

Much as I regret it, we shall now have to proceed with operation 'Illusionist'. The outline of the scheme is given below and the Deputy Engineer, who has carried out a similar operation in the past with gusto, will fill you in with all the details. Briefly the operation will run as follows.

The Resident Engineer will declare that his birthday is on Friday, 8 September and will invite the whole of the gang and staff to a local hostelry for a lunch-time drink at 1 pm. He must be on good terms with Jenkins by this time and Jenkins will be included in the invitation. The trench must be bottomed up and all pipes prepared for laying and jointing. The Deepwet reservoir heavy duty gang, with you in charge, must be ready and waiting in their vehicles around the corner at 12.55 pm for the site to be cleared.

The resident engineer and all the men will not return before 2 pm by which time you should be on the way back to Deepwet. I assume that the pipes will then have been laid and, if your conscience is troubling you, you can flatten it further by working out a method of recovering the Resident Engineer's expenses to the satisfaction of our auditor.

Civil Engineer's Office
Willpipe Central Station
Engineer and Manager *11 September*
Water Board Offices

Dear Sir,
Two Arches Rail Crossing
I must protest at the way work was carried out at Two Arches last Friday. The 24 in. main was positioned in the trench while my Inspector of Works was temporarily absent from the site for personal reasons.
Please inform me how the pipes were installed in the trench in compliance with British Rail Conditions.

Yours faithfully,
S.T. Standfast,
Divisional Civil Engineer

Water Board Offices
Divisional Civil Engineer 13 September
British Rail

Dear Sir,
Two Arches Rail Crossing
I acknowledge your letter of the 11 September in connection with the above but I regret that I cannot assist you in this matter.

I appreciate your difficulties but may I remind you that I was requested to solve my own problems. I now consider that an explanation of the successful solution is your problem.

Yours faithfully,
Engineer and Manager

5

J.S. Diggleby Esq
Hon. Secretary
Belfordshire Archaeological Society

Dear Sir,

Wellgreen Reservoir Scheme

You may recall that your Society formally objected at the Wellgreen Public Enquiry in connection with the probable inundation of the 17th century farmhouse and provision was therefore made for the farmhouse to be carefully dismantled and re-erected on a more appropriate site made available by your society.

I now write to say that the Contractor is ready to begin this work and I should be grateful if you would let me have details of the new site so that the necessary arrangements may be put in hand.

Yours faithfully,
Engineer and Manager

Hon. Secretary
Belfordshire Archaeological Society
Engineer and Manager   *6 September*
Willpipe Water Board

Dear Sir,

Wellgreen Reservoir Scheme

With reference to your letter of 22 August, I can now inform you that further investigation by members of this Society reveals that there is so much apparent restoration work on the farmhouse that it is not worth retaining. However, it has been unanimously agreed that the main item of archaeological

17

interest is the fine specimen of a Dutch oven in the farmhouse and I am instructed to say that this should be carefully preserved for exhibition in a local museum.

Yours faithfully,
J.S. Diggleby

Water Board Offices
8 September

J.S. Diggleby Esq
Hon. Secretary
Belfordshire Archaeological Society

Dear Sir,

Wellgreen Reservoir Scheme

Thank you for your letter of 6 September. The Dutch oven can easily be extracted, transported and re-erected. I should be grateful therefore if you let me know which museum wishes to acquire this exhibit.

Yours faithfully,
Engineer and Manager

Hon. Secretary
Belfordshire Archaeological Society
21 September

Engineer and Manager
Willpipe Water Board

Dear Sir,

Wellgreen Reservoir Scheme

I regret to inform you that none of the local museums wish to acquire the Dutch oven from the 17th century farmhouse. The consensus of official opinion is that the oven is of dubious antiquity and that the county abounds in numerous and better examples.

The Society therefore renounces all claims to the farmhouse and I trust that no inconvenience has been caused.

Yours faithfully,
J.S. Diggleby

18

The Bellfordshire Archaeological Society has now retired from the scene

INTERNAL MEMORANDUM
23 September
To: Resident Engr, Wellgreen From: Engineer and Manager

You will be glad to know that the Belfordshire Archaeological Society, despite all their impassioned pleading at the Public Enquiry, have now withdrawn from the scene.

   That damned ruin in field OS 274 can now be demolished and I would suggest that the hardcore be bulldozed out and used as bottoming for the car park areas.

INTERNAL MEMORANDUM
28 September
To: Engineer and Manager From: Resident Engr, Wellgreen

You will no doubt be interested in the contents of the attached parcel. It was turned up by the dozer engaged on demolishing and moving the farmhouse and must have been somewhere under the foundations.

   You will have to thank the dozer driver (P.D. O'Neill) for collecting it. He paid it particular attention because it projected over the dozer blade and he first thought it was a crucifix! Incidentally I never knew they used rivets in those days!

Keeper of Archaeology
British Museum
6 October

Engineer and Manager
Willpipe Water Board

Dear Mr Fitzjohn,

   On behalf of the Museum, I have great pleasure in accepting an exhibit from the Water Board.

   The Board must have taken great care in carrying out the excavations at Wellgreen to avoid causing damage to such an interesting find. It will go on permanent exhibition immediately as it is probably the finest example of a Bronze Age sword in the country.

Yours sincerely,
J.E. Appleyard

6

Meteorological Office
23 June

Engineer and Manager
Willpipe Water Board

Dear Sir,

I should be grateful if you would confirm and check the rainfall readings from Deepwet gauge No.7. During wet weather they are consistently higher than others in the locality, as the attached figures show. Has the location or environs of the gauge changed recently.

Yours faithfully,
M.J. Newman (Miss)

Water Board Offices
26 June

Miss M.J. Newman
Meterological Office

Dear Madam,

Deepwet Gauge No.7

Thank you for your letter of 23 June. The readings from the above gauge are correct. They are taken by the reservoir keeper, Mr Brown, who attended one your 'Observers' courses two years ago and the location of the gauge has not changed since it was agreed with Dr Glasspoole when it was first installed.

Nevertheless we shall give the gauge some additional attention in future in view of your remarks.

Yours faithfully,
Engineer and Manager

INTERNAL MEMORANDUM
26 June

To: Supplies Engineer   From: Engineer and Manager
I attach copies of correspondence with the Meteorological

Been venting his spite by adding to the contents of the gauge

Office. If we assume the readings are correct it might mean that the reliable yield of this catchment is higher than the accepted figure and we need every gallon from Deepwet reservoir while we are waiting for completion of the Wellgreen scheme.

Please check the correlation of No.7 gauge with the main stream inlet weir recorder and the runoff figures for the reservoir for the last few years.

INTERNAL MEMORANDUM
28 June
To: Engineer and Manager   *From: Supplies Engineer*

There's certainly a mystery attached to the Deepwet statistics.
No.7 gauge started giving higher readings than normal about two years ago and these readings correlate remarkably accurately with the charts from the weir recorder at Murrain Beck. However, the reservoir runoff figures are unchanged and it looks as if we are losing water somewhere. Perhaps a swallow hole has opened where the limestone outcrops at the upstream end of the reservoir? Investigations are continuing.

INTERNAL MEMORANDUM
29 June
To: Supplies Engineer   From: Engineer and Manager

With reference to the Deepwet runoff, I have a lot of faith in impounding reservoirs and a healthy bump of scepticism regarding instruments. Please check the latter first.

INTERNAL MEMORANDUM
5 July
To: Engineer and Manager   *From: Supplies Engineer*

I'm pleased to report that the Deepwet mystery has been solved.
As it was a fine weekend, my wife and I took a walk around Deepwet reservoir and through Murrain Wood. I thought I'd

take the opportunity to look at the weir recorder as it is at least a mile from the road and I have not inspected it for years. I found the recorder house in a dilapidated state, the windows and doors being covered with cobwebs and the lock rusted solid.

Brown reported to my office this morning and confessed that, ever since he took over from Mr Rainchart two years ago, he has avoided the weekly walk to the recorder house and has faked the recorder charts, basing them upon previous records and the readings obtained from Rainfall Gauge No. 7. He swore that he had faithfully taken the rainfall readings because he needed them for his other skullduggery. However, further enquiries have revealed that the village poacher, who rejoices in the name of Lapwing and is renowned for being a writer of anonymous letters as well as a sworn enemy of Brown, has been venting his spite by adding to the contents of the gauge when it is raining. He says that he poured water into the gauge from a 'liddle bottle' but I would bet on micturition.

I hope you will confirm my action with regard to Brown. I told him he would be suspended for a fortnight without pay and then he would be transferred to the Work Study Section.

7

Willpipe General Hospital
Casualty Department
14 June

Water Engineer and Manager
Willpipe Water Board

Dear Sir,

I am writing to you on the subject of industrial accidents because, on many occasions recently, I have examined your employees on admittance to the hospital. The cases I refer to are:

(1) Mr D.B. (Filter attendant): Chlorine poisoning
(2) Mr F.O. (Reservoir keeper): Multiple lacerations
(3) Mr E.T. (Labourer): Broken fingers
(4) Mr K.S. (Plumber): Broken leg, concussion and pneumonia
(5) Mr T.W. (Foreman): Dislocated shoulder

I would suggest that your safety regulations need revising and I should welcome your suggestions.

Yours faithfully,
I.M.N. Parker,
Senior House Physician

Willpipe Water Board
16 June

Dr I.M.N. Parker
Senior House Physician
Casualty Department
Willpipe General Hospital

Dear Dr Parker,

I appreciate your civic and medical conscience regarding the accidents mentioned in your letter of the 14 June and I feel you deserve some additional information for your case histories. Accordingly, my observations are as follows.

(1) *Mr D.B. (filter attendant)*

Part of D.B's duties is to maintain the chlorination equipment at Silverstone which is an old, but still efficient, Wallace and Tiernan machine with a bell-jar on a pedestal. The bell-jar is removed and cleaned once a fortnight by a safe procedure, the attendant wearing a mask in case there is a leakage of gas. Mr D. B. suffers from asthma and, for some unknown reason, became convinced that a whiff of chlorine could remedy this condition. Accordingly he removed the bell-jar without either closing down the machine or wearing the mask and inhaled deeply. I understand that, under your care, he should be fit to resume work and I should be interested to know whether he is still asthmatical?

(2) *Mr F.O. (reservoir-keeper)*

While digging a trial hole near the shoulder of the embankment, Mr F.O. uncovered the end of a disused 18 in. dia. pipe. Instead of making enquiries at head office, he decided to investigate by crawling up it, but in case he could not crawl backwards he attached a rope to his waist and left instructions with two labourers to haul him out if he got stuck and gave two jerks on the rope. Unfortunately he forgot to say 'slowly and gently' and when he eventually gave the signal his men pulled him out with such enthusiasm that the internal nodules removed large areas of clothes and skin.

(3) *Mr E.T. (labourer)*

Mr E.T. was granted leave of absence to attend at his dentist's surgery for an extraction. Mr E.T. is a person who refuses the aid of an anaesthetic (general or local) and instead takes the strain by entwining his fingers round a 6 in. nail. In this case the nail proved stronger than his fingers.

(4) *Mr K.S. (plumber)*

Mr K.S. has the reputation of being the Casanova of Willpipe City. Several weeks ago he was called out on a wet and windy night to carry out an emergency repair at a consumer's house. The man of the house was on night shift, the lady was attractive, willing and in a night shift and Mr K.S. succumbed, throwing discretion to the winds. The husband returned unexpectedly and Mr K.S. beat a hasty retreat through the bedroom window clad only in a wristwatch and fell into an ornamental pond.

Beat a hasty retreat through his bedroom window clad only in a wrist-watch

(5) *Mr T.W. (foreman)*

Mr T.W. is gullible and bald. While supervising the installation of a cross-country main, the route passing alongside a gipsy encampment, he took his troubles to the matriarch of the tribe and eventually purchased from her a vile-smelling concoction which was reputed to possess hair-restoring properties. He poured this liquid into a bowl on the site (his wife has a sharp and sarcastic tongue) and immersed his head in it. A car backfired in the vicinity and startled him. In this awkward posture he lost his balance, fell into the trench and dislocated his shoulder. I might mention that he is still bald but has the hairiest ears in the county.

If you are still of the opinion, my dear doctor, that the Board's safety regulations could be rewritten to prevent the above calamities, I should be pleased to have your assistance.

Yours sincerely,
Engineer and Manager

8

The Laurels
Bellford Common
8 December

Engineer and Manager
Willpipe Water Board

Dear Sir,

My great-uncle died recently and was cremated. In his will he expressed a wish that his ashes be scattered from a boat in the centre of the Deepwet River.

May I have the permission of the Board to perform this act on Saturday, 18 December.

Yours faithfully,
P.J. Littorall

Willpipe Water Board
9 December

P.J. Littorall Esq
The Laurels
Bellford Common

Dear Sir,

With reference to your letter of the 8 December, I fear that you have approached the wrong organization. I would suggest that you apply to the Bellfordshire River Authority, who can solve your problem.

I might mention that, as they are the Licensing Authority, the Board even has to apply to them for a licence when we wish to abstract water from the river.

Yours faithfully,
Engineer and Manager

The Laurels
Bellford Common
10 December

Engineer and Manager
Willpipe Water Board

Dear Sir,

I object to being passed from one organization to another in connection with a simple request. I am not a fool and I naturally wrote to the River Authority in the first place. I was informed that your Board is the statutory water undertaker for the area and I must insist that you now give me a straightforward answer to my application.

Yours faithfully,
P.J.Littorall

John P. Aitkin LLB
Clerk to Bellfordshire River Authority
Engineer and Manager   *15 December*
Willpipe Water Board

Dear Robert,

What sort of games are you playing with your consumers? I received a cheque and an application for a colour TV licence from a Mr Littorall and when I enquired why he had sent it to me, he said that you had assured him that we were the Licensing Authority. Is he an eccentric?

Yours truly
John

Willpipe Water Board
16 December

P.J. Littorall Esq
The Laurels,
Bellford Common

Dear Mr Littorall,

I should very much like to make your acquaintance and may I invite you to luncheon with me on

A wish that his ashes be scattered from a boat in the centre of the
Deepwet River

Tuesday, 21 December. Perhaps you would be good enough to meet me at my office at 12.15 when we can discuss the disposal of your great-uncle.

Yours sincerely,
Engineer and Manager

Willpipe Water Board

John P. Aitkin Esq
22 December

Clerk to Bellfordshire River Authority

Dear John,

I am now in a position to reply to your letter of 15 December. Many years ago, I learnt that the only way to deal with the Littoralls of this world is on a personal friendly basis and I had a most enjoyable lunch with the gentleman yesterday.

He is an eccentric and a self-made one. He gets considerable fun out of crazy correspondence with people of our ilk though I hope he has now put the water board behind him. During our conversation I found out that he had sent a gold hunter for repair to the Bellford Watch Committee; he applied for planning permission for an awning to the Blind Institution; he applied to the Marriage Guidance Council for a qualified instructor and an introduction to a rich spinster, he offered to invest £20 a month in the unit funds of the National Trust and sent an order for 1 dozen rose bushes to the Institution of Plant Engineers. When leaving the restaurant we overhead a cliché from the adjoining table about something being 'welded into an organic whole'. At this, his eye gleamed and he mentioned that he would be writing to Institution of Mechanical Engineers and the Institute of Biology. Incidentally, while thanking me for my hospitality, he added that this was no doubt a prerequisite of my qualification!

I suggest you send him his TV licence!

Yours sincerely,
Robert

9

Barchester and District Water Co.

Engineer and Manager
Willpipe Water Board

9 February

Dear Fitzjohn,

Mr S.C. Rounger

The above named, who I believe to be an engineering assistant on your staff, has applied for the appointment of Assistant Distribution Engineer with my Company and has quoted your name as a referee. He will be interviewed by myself and the Managing Director on Wednesday, 24 February.

I should be grateful if you would give me your confidential opinion as to his character, conduct and engineering ability. I should like to know whether you consider him capable of supervising two district offices and about 50 men.

Your sincerely,
A.J. Griddle,
Chief Engineer

INTERNAL MEMORANDUM
10 February

To: Distribution Engineer   From: Engineer and Manager

Mr S.C. Rounger

I have been asked to supply a reference for Mr S.C. Rounger. He has only been with us for 12 months and I have had little to do with him since last August when I had him on the carpet for that disgraceful affair at Silverstone. Has he changed his tune recently?

I have to comment on his character, conduct, engineering ability and suitablity for supervising about 50 men.

INTERNAL MEMORANDUM
12 February

To: Engineer and Manager   From: Distribution Engineer

With reference to your memo of last Wednesday, I have been compiling a dossier on Mr Rounger and was hoping to present you with sufficient evidence at the end of the month to warrant his dismissal although I realise that this might prove to be extremely difficult. We were certainly conned when we interviewed him for the appointment originally.

He is scruffy, idle and irresponsible. He cannot be trusted with any major works and generally mismanages the few minor matters that he cannot avoid. I would not consider him suitable to supervise a wheelbarrow. I just do not understand him.

Engineer and Manager
Willpipe Water Board
Chief Engineer   *15 February*
Barchester and District Water Co.

Dear Griddle,

Mr S.C. Rounger

On the understanding that this is a confidential letter and will be seen by your eyes only, I can tell you that Mr Rounger, who has been with me for 12 months, is a disgrace to his profession and I would welcome the opportunity of getting rid him.

He made a complete hash of the one important job entrusted to him and, being naturally idle, has tackled few minor matters since, generally in an unsatisfactory manner. He is scruffy, idle and irresponsible; does not mix well with his colleagues and is not able to supervise one man let alone 50.

I cannot recommend him to you.

Yours sincerely,
Engineer and Manager

INTERNAL MEMORANDUM
26 February

To: Engineer and Manager   From: S.C. Rounger

Thank you for giving me a reference for the appointment with

I attended for the interview last Wednesday and have been offered and accepted the post

Barchester and District Water Co. I attended for interview last Wednesday and have now been offered and accepted the post.

I therefore wish to tender my resignation and ask you to accept one month's notice terminating on 31 March.

Barchester and District Water Co.
9 June

Engineer and Manager
Willpipe Water Board

Dear Sir,

Mr S.C. Rounger

I must firmly protest about the outrageous trick you have perpetrated upon my Company. When I received your reference in connection with Mr Rounger, I could not understand it. However, knowing of your literary pursuits I came to the opinion that you were indulging in hyperbole and were using this method because you wished to retain his services and did not wish him to transfer to this Company. His performance at the interview reinforced this opinion.

Unfortunately I find that I have been wholly deceived and that your letter contained the complete and utter truth. One normally damns with faint praise and I have never before in 30 years experience encountered such a dangerous document. I consider that it comes close to being a breach of professional ethics and it is only the fact that it was a confidential letter that prevents me from reporting you to the President of the Institution.

10

OFFICE MEMORANDUM

To: All staff   From: Engineer and Manager

In view of the number of letters and reports that need to be sent back for retyping and redrafting, I make no apology for mounting my hobby-horse again and re-issuing a memorandum that was circulated about five years ago. The most popular mistakes are listed below and, if they are avoided, documents will be read, understood, filed and found more easily. Some of the points are absurdly elementary but they nevertheless recur on many occasions.

(1)   Quote references on all letters.

(2)   Do not write one letter on two different subjects. (Letters are filed by subject not addressees' names.)

(3)   An individual may be addressed as 'Dear Sir' but a firm is a plural entity.

(4)   Please avoid archaic introductions such as 'Yours of the 26th ult. to hand.'

(5)   Do not hesitate to consult a good dictionary to confirm the correct spelling of a word. Incidentally, my test for a good dictionary is that it contains the words 'pejorative' and 'subdolous'.

(6)   All letters and reports should be correctly punctuated.

(7)   All sentences should be kept short and not allowed to flow painfully down the page fortified with liberal injections of the word 'and'.

(8)   Avoid the use of abbreviations (especially initials) as the reader may not recognize them. Initials may be used if preceded by the full title earlier in the text.

(9)   Please use the correct case for the first person singular. I often see a phrase like 'The Treasurer asked the engineer and I to provide...' when the writer would never dream of saying 'The Treasurer asked I to provide...'.

'Take a letter Miss Smithers...'

(10)   Please avoid ending a sentence with a preposition.

(11)   Whereas the direct double negative is never seen, the indirect double negative appears to exert a fatal fascination. The phrase 'I am not unaware that...' should only be used to convey a subtle shade of meaning.

(12)   Please ensure that the word 'different' is followed by the preposition 'from' and do not say 'different than' or 'different to'.

(13)   Remember that one cannot have 'the best' of two alternatives but only 'the better'.

(14)   Beware of the concealed modifier. A sentence such as 'In his fortieth year of service, Mr Brown's proud boast was that he had never lost a day due to sickness' indicates that not only is the boast masculine but also remarkably healthy.

(15)   Please eschew the use of unnecessary introductory phrases such as 'I make no apology for...' or 'I have no hesitation in saying that '. When I read these particular examples, I get the impression that the writer is either apologizing or hesitating.

(16)   Please avoid the use of polysyllabic jargon when simple words would suffiice.

(17)   Please avoid the phrase 'to forward plan' which actually means 'to arrange beforehand beforehand'.

(18)   Please avoid hackneyed journalistic words like 'probe', 'contact' or 'viable' unless you are certain that they mean exactly what you intend them to mean. Another absurd expression is 'He has come all the way from...'. If he had not, he would not have arrived!

(19)   Finally I ask for consistency in the use of tenses of verbs. While on the subject of verbs, I would mention that I never know whether to say 'none is' or 'none are'. To quote Mr Richard Mallett, if one automatically assumes that 'none' is singular then any alteration to the saying 'None but the brave deserve the fair' indicates an unwarranted concern for the love life of a Red Indian.

If any staff member is offended by the above points, he or she can get revenge by replying and pointing out my foibles or mistakes for none of us is (?) perfect.

A type of resonance was set up due to the action of the pneumatic hammer

11

INTERNAL MEMORANDUM

To: Engineer and Manager   From: New Works Engineer

A leak was detected at Doomhead service reservoir and excavation through the embankment revealed that there was a crack in the 12 in. outlet pipe from No.1 compartment where it passes through the RC wall upstream of the flexible joint. It has been arranged to empty this compartment next Monday, cut through the wall using pneumatic hammers, replace the pipe and re-concrete the wall.

INTERNAL MEMORANDUM

To: Engineer and Manager   From: New Works Engineer

This is a report to confirm our discussion this morning and subsequent inspection at the service reservoir.

When the maintenance gang completed cutting the hole through the wall of Doomhead service reservoir, the foreman observed that 41 of the 49 columns in the tank were cracked and the concrete had spalled off the edges of the columns facing towards the centre of the reservoir. The reservoir had previously been inspected 15 months ago when no defects were visible.

You and I carried out an inspection of the affected compartment and also of the adjacent compartment which had been emptied in readiness. The damage in No. 1 compartment was confirmed but No. 2 compartment proved to be in perfect condition.

Arrangements have been made to erect temporary props to hold the roof, samples of steel and concrete will be obtained for tensile and compressive tests and an independent firm of consulting engineers will be requested to carry out a complete investigation.

Willpipe Water Board

Messrs Standfast, Coolhead and Partners
Consulting Engineers

Dear Sirs,
        In confirmation of our telephone conversation, I should be grateful if your firm would carry out an immediate investigation of an inexplicable occurrence at Doomhead service reservoir. This is an underground reinforced concrete service reservoir, consisting of twin compartments each holding one million gallons. It was designed in this office, constructed by contract and has been in commission for three years without incident. Details of the construction and the defects are attached in a separate report and the Board would wish you to check the design and construction and report upon the reason for the cracked columns. My office will provide whatever assistance and co-operation you require.

Yours faithfully,
Engineer and Manager

Standfast, Coolhead and Partners
Consulting Engineers

Engineer and Manager
Willpipe Water Board

Dear Sir,

Doomhead Service Reservoir

We have now completed our commission consisting of an investigation into the cause of the cracked columns in the above service reservoir and our detailed report is attached. A summary of our findings is as follows:

(1)   The design is in accordance with CP 2007 and cannot be faulted.
(2)   The construction of the reservoir was carried out competently and was well supervized by an experienced resident engineer. The standard of workmanship is very high and examination of site diaries shows that no untoward incidents occurred during the construction of the floor, columns and roof.

42

(3)   A survey of the reservoir confirmed that construction was carried out to specification.

(4)   Tests carried out on samples of steel and concrete gave excellent results and analysis of concrete specimens showed that the mix was as specified.

(5)   Consideration was given to stresses set up not only by applied loadings but also by other causes such as shrinkage, creep, temperature movements, foundation movements, external and upward water pressure. Thirty equations of equilibrium combined with nine different sets of loading or movement conditions were postulated and solved on a computer. The solutions definitely indicated that the loadings applied would not cause movements sufficient to cause column failure and the normal design methods used were adequate and accurate.

(6)   Advice has been sought from, and the reservoir inspected by, engineering geologists, soil mechanics consultants, research associations, industrial research laboratories, universities, other consulting engineers and the War Office. Without exception, no one has been able to offer a satisfactory explanation and this type of failure appears to be outside existing experience.

Consequently, as has been said in other places, when all probable explanations have been examined and found wanting, what remains - no matter how improbable - must be the answer.

We therefore consider that a type of resonance was set up in the whole structure, but particularly in the columns, due to the action of the pneumatic hammer. This resonance caused repeated rapid flexure of the columns and final cracking and shattering. The frequency of vibration of the hammer is 1000 per minute and we have calculated the natural frequency of vibration of the roof slab and columns to lie within the range 900-1100 per minute.

As a possible check on this theory, a variable speed DC motor with an offset flywheel was strapped to a column and the column appeared to resonate when the motor was running at about 1000 rpm. You will appreciate that it would only be possible to confirm this hypothesis absolutely by carrying out a similar exercise in No.2 compartment!

We can, however, definitely state that no blame can be attached to the Water Board or the Contractors for this failure.

Yours faithfully,
Standfast, Coolhead and Partners

INTERNAL MEMORANDUM
To: New Works Engineer   From: Engineer and Manager

I attach, for your information, a letter and report from the consulting engineers investigating Doomhead reservoir.

I believe that they have come to the correct conclusion but we have now attained the dubious distinction of possessing a unique explanation which no one (unless he has been intimately concerned with the affair) will believe, apart from our Insurers who will be able to deny liability under the terms of the policy.

The columns should now be recast and I would suggest that each one incorporates a hard rubber section with a thick stainless steel dowel for continuity. I would also say that, if any RC structural walls have to be cut in future, it would be prudent to use an oxygen lance.

12

INTERNAL MEMORANDUM

To: Distribution Engineer   From: Engineer and Manager

The Board has now decided to abolish all charges for hosepipes used for watering gardens and washing cars providing that they are hand-held during operation. However it has been decided to introduce a charge of £5 per annum for automatic sprinklers or fixed hose pipes. The revenue will be negligible but it is hoped that a charge of this nature will dissuade consumers from using these appliances which waste considerable quantities of water and can cause havoc with distribution pressures on new estates.

Builders' merchants and plumbers will inform us regarding purchasers and your inspectorate staff should be especially vigilant to detect the use of any unauthorized sprinklers. We cannot cover all sources of supply but the word will get around about our inspections.

Beech Hedge Cottage
The Manager   *Cloughstone*
Willpipe Water Board

Dear Sir,

I must protest in the strongest terms regarding the violation of privacy committed by one of your inspectors. Last Thursday, without due notice, he entered the rear garden of my property which is completely enclosed by a high beech hedge, and surprised my wife who was sunbathing in the nude. My wife was alone on the premises at the time and the shock was so great that she is now in a highly nervous state. I intend to consult my solicitor and obtain compensation for this outrage. It was also grossly impertinent of the fellow to announce his presence by saying 'Good afternoon, sir'.

Yours,
M.J. Pompass

My wife was alone on the premises at the time and the shock was so great that she is in a highly nervous state

INTERNAL MEMORANDUM
To: Distribution Engineer   From: Engineer and Manager

I enclose a copy of a letter from a Mr M.J. Pompass. Please let me have details of the incident to enable me to reply.

INTERNAL MEMORANDUM
To: Engineer and Manager   From: Distribution Engineer

The inspector concerned with the incident at Beech Hedge Cottage was Jack Brown. He was following up the sale of an automatic sprinkler to Mr Pompass who has not paid for a licence. The sprinkler was in operation that afternoon and Brown followed the usual procedure. He continually amazes me with his talents.

Willpipe Water Board

M.J. Pompass Esq.
Beech Hedge Cottage
Cloughstone

Dear Sir,

      I have your letter of complaint regarding one of the Board's inspectors - Mr J. Brown. Last Thursday afternoon realising that an automatic sprinkler was in operation in your back garden, he knocked at your front door and received no reply. As he has a right to enter premises within the area of the Board at all reasonable hours for the purpose of his official duties (Section 82, Third Schedule, Water Act 1945), he made his way to the back garden to speak to the occupant of the house. As his appearance caused distress to Mrs Pompass he withdrew immediately without completing his duties and I accordingly take this opportunity of enclosing an account for £5 which will permit you to continue to use the automatic sprinkler.

      I might mention that Mr Brown is completely blind but his hearing is so acute that he is invaluable in the detection of leakages and the use of unauthorized appliances.

Yours faithfully,
Engineer and Manager

**A gang of workmen running along the road overtaking each other
and resting their head on a stick**

13

Engineer and Manager
Willpipe Water Board

Members' Room
Willpipe City Hall

Dear Mr Fitzjohn,
      You may remember I attended my first meeting as a new member of the Water Board yesterday and found myself a little at sea when the various works and schemes were being discussed. I should like to make myself more familiar with the Board's affairs and I should be grateful if you would provide me with a permit to visit the reservoirs, treatment works and pumping stations. I do not want a conducted tour but would prefer to carry out a personal inspection.

Yours sincerely,
J.P. Greenwan,
Councillor

Willpipe Water Board

Dear Councillor Greenwan,
      As requested, I enclose your permit herewith. I can assure you that you would find it advantageous to be accompanied on your visits but I will not press the point - if you should change your mind, please let me know.

Your sincerely,
Engineer and Manager

Engineer and Manager
Willpipe Water Board

Members' Room
Willpipe City Hall

Dear Mr Fitzjohn,
      I have visited several of the Board's install-

ations and, as I expected, these unheralded visits enabled me to uncover several matters which would not have been apparent if I had been taken round by one of your assistants.

I am appalled by the laxity of your employees and the frivolous way in which the supervisors answered my questions. I intend to raise the following points at the next Board Meeting:

(1)   On my way to Deepwet reservoir, I came across a gang of workmen running along the road overtaking each other and then resting their heads on sticks. Luckily there was no traffic or they would have been seriously injured. I tried to speak to one of them but he said 'no time mate, come back tomorrow'.

(2)   At the high level pumping station, I saw two fitters dismantling a pump. One was attempting to undo a screw with a hammer while the other was putting some oil on a nut and talking to it; he called it 'Pierre'. The men were obviously mentally unbalanced and I left hurriedly without speaking to them.

(3)   At Deepwet treatment works, I was looking at the chart on the outlet meter and I noticed that the flow had increased rapidly at one point. I asked a man, who was passing, what was the reason for this and, over his shoulder he said 'That marks the end of the Morecambe and Wise Show'. Does he think I'm a fool?

(4)   Outside the treatment works, I saw a glass ball on a stand. I asked the same man what it was for and when he started to say 'That's a crystal ball for estimating...', I cut him short and told him I wanted no more frivolity.

I doubt that there can be any explanation for these incidents but I am willing to hear your excuses.

Yours sincerely,
J.P. Greenwan

Willpipe Water Board
Dear Councillor Greeenwan,
           I have your letter of complaint and while excuses are unnecessary, I will try to clear up your misunderstandings.

(1)   The workmen on the road were reconditioning a 15 in. main. This is normally done by forcing a scraper through the main by hydraulic pressure. The nose of the scraper carries a radio-active tracer which can be detected by what you would call a Geiger counter in case it becomes blocked in any position. Unfortunately the detector was not working and the men improvized by listening to the noise of the scraper in the main as it progressed along the road. They used pick handles as listening rods.

(2)   With reference to the two fitters at the pumping station, I can assure you that a sharp blow with a hammer and punch is an accepted method of 'starting' a recalcitrant slot-headed bolt. The other fitter belongs to an esoteric religious sect which believes that all inanimate objects have souls. Who am I to deny this belief which has the effect of making him the most careful and reliable fitter I have yet encountered.

(3)   The answer to your query about the Deepwet outlet meter was factually correct. All flow meters indicate a sharp rise in water consumption at the end of a nationally popular television show. I am sure you can deduce the reason.

(4)   In connection with your final query, I'm afraid you were impulsive in interrupting the man's answer. He was about to say 'That's a crystal ball for estimating the hours of sunshine'. It is one of the instruments in our climatological station and the ball focusses the rays of the sun to burn a trace on a measuring card.

I trust that you now appreciate that it would be preferable if you were accompanied on future inspections.

Yours sincerely,
Engineer and Manager

51

Visitors included two television crews, a radio team, 19 newspaper and magazine journalists with accompanying photographers...

14

INTERNAL MEMORANDUM
5 April

To: Supplies Engineer   From: Divisional Manager

I attach, for your information, a copy of a letter dated 4 April from our old friend, Mr P.J. Littorall. You may recall our brush with him a couple of years ago when he required us to distribute the ashes of his dead uncle in Deepwet River because we were the statutory water undertakers in the area. He is an innocent practical joker who specializes in causing baffled embarrassment in his victims.

I suggest you read the Willpipe Express carefully during the next few days and be prepared for some extraordinary incidents at Deepwet Reservoir.

The Laurels
Bellford Common

The Divisional Manager   *4 April*
Willpipe Water Division

Dear Mr Fitzjohn,

We have neither met nor corresponded since we had that pleasant luncheon together nearly three years ago and I decided to remedy this omission.

I have been reading with interest the various letters published in the Willpipe Express from fishermen complaining that there are not sufficient boats for hire at Deepwet Reservoir. My contribution on an associated subject should be printed shortly and you may find it a welcome relief from routine work.

Yours sincerely,
P.J. Littorall

Excerpt from 'Letters to the Editor' in the issue of the *Willpipe Express* dated 8 April

Sir,
   While walking along the footpath to the east of Deepwet Reservoir on Thursday, I heard a splash in the water and, peering through the twilight, I discerned a large creature swimming towards the embankment and thrashing the surface with its tail. It seemed similar in shape to the drawings I have seen of the Loch Ness monster and was at least 80 feet long. My friends tell me that it must have been an hallucination but I cannot accept this explanation although my eyesight is not as good as it used to be.
   I realise that Deepwet is a man-made reservoir but there have always been large eels in that water and, with all the radioactive nuclides (particularly radiocaesium) that were deposited in inland waters from rainfall following nuclear test explosions in the early sixties, it is not impossible that one species has mutated to produce a giant.
   I should be interested to know whether any other readers have observed this phenomenon at Deepwet reservoir. Complaints have been made about a shortage of hire boats for fishermen - is this due to boats being destroyed by this creature and perhaps attributed to vandalism by the waterworks authorities?

Yours, etc.
P.J. Littorall

(Editor's note: The Manager of the Willpipe Water Division was asked for his comments and he replied that it was amusing but fantastic nonsense).

Willpipe Naturalists Society
43 High Street
Willpipe
9 April

The Manager
Willpipe Water Division

Dear Sir,
   I note that you describe the letter from Mr P.J. Littorall

in the Willpipe Express yesterday as amusing but fantastic non-sense. Let me say at once that I can confirm his story. I was stationed at Deepwet all day on Sunday carrying out a wild fowl count and decided to return home at dusk about 45 minutes after sunset. I was looking towards the upstream end of the reservoir when I saw a series of humps moving through the water towards the west bank. There was a uniformed man, perhaps a bailiff, on the bank at this point and I conjectured that this creature, which must be the same one observed by Mr Littorall, was going to attack him. The man promptly ran away into the trees and, al-though he was some distance away, I ran towards his location in order to render assistance. When I arrived, both the man and the Deepwet monster had disappeared.

My nephew is a journalist employed by the Willpipe Express and I intend to tell him my story for publication in the paper. A large scale investigation should begin at once so that the creature can be caught and studied.

Yours faithfully,
A.D. Plover,
Hon. Secretary

INTERNAL MEMORANDUM
10 April

To: Supplies Engineer   From: Divisional Manager
I attach a copy of another letter from Mr A.D. Plover in the monster saga and I asssume that it will be splashed all over the local paper tonight. Mr Littorall will have a lot to answer for in the future but in the meantime you will have your hands full at Deepwet over the Easter Holiday. You had better arrange for additional patrols and stand-by men to assist you at overtime rates over the weekend. Please let me have a full report next Wednesday, including the bailiff's story.

INTERNAL MEMORANDUM
17 April

To: Divisional Manager   From: Supplies Engineer
First of all, let me give you the explanation of Mr Plover's

adventure. (Mr Littorall just possesses a fertile imagination).

At the conclusion of sailing on Sunday 7 April, the bailiff on duty noticed that two of the roped buoys that delineate the reservoir shallows had been punctured and were sinking. Consequently, before he went off duty, he untied the rope on the east bank of the reservoir, walked around to the west bank, again untied the rope and then pulled the line of buoys across the reservoir, up the bank and into the stores hut in the trees. He then went home leaving the 'monster' to be attended to next day.

Next, the incidents at Deepwet from Friday 12 April to Tuesday 16 April. Over these five days, visitors to Deepwet included two television crews, a radio team, nineteen newspaper and magazine journalists with accompanying photographers and no less than 23 000 casual sightseers. The usual fishermen and sailors could not drive near the place and anyone walking to the reservoir soon gave up in disgust. Nearly all the general public arrived by car and traffic jams lasted each day until 10.30 pm. There were 18 car accidents (nobody hurt) and 12 cars broke down. The local garages, AA and RAC men were busy from dawn to sunset and the police superintendent had to station 20 policemen and 3 patrol cars at the reservoir each day. Also present were the usual commercial interests consisting of ice-cream vendors, hot-dog salesmen and mobile fish and chip shops.

The hire boats were engaged by television and newspaper agencies for the first two days but were idle for the remainder of the holiday as no one else wished to venture into deep water. I assume that they will be under-used for the rest of the season. Acting under normal regulations, no one was allowed inside the perimeter fence unless he or she was in possession of a daily fishing licence (£1 + VAT). I am therefore pleased to report that our income over the five days amounted to £14 718 and this will obviously be the first year that we have ever made a profit on our amenities.

15

Willpipe University
Civil Engineering Dept
15 April

R. Fitzjohn Esq.
Willpipe Water Division

Dear Mr Fitzjohn,

As you have purchased a ticket, you will be aware that the Annual Dinner of the Institution's West Central Association will be held at the George Hotel in Willpipe on Friday, 19 April. The principal speaker, giving the Toast of the Institution, will be Dr Phineas Barnum, Director of the Chilton Hundreds Management College. The reply was scheduled to have been given by the President of the Institution, who is a personal friend of Dr Barnum. Unfortunately, as you may have heard, the President broke his leg on Saturday (a simple fracture) but he is now in plaster and will not be able to make the journey to Willpipe. His place will be taken by the Chairman of the West Central Association who had intended to propose the toast to the guests and we are therefore faced with a vacancy in this latter position.

By some unfortunate coincidence, neither of the vice-chairmen are available on that evening; one being confined to bed with pneumonia and the other being in the West Indies. The past chairman is in America and you appear to be the most senior of the Committee members (although we do not seem to have had the pleasure of your company at many committee meetings recently.) I should therefore be most grateful if you would step into the breach. You will, of course be refunded the cost of your ticket.

Yours faithfully,
C.B. Whistfall,
Honorary Secretary,
West Central Association, ICE

Took the opportunity to ride his hobby-horse

Willpipe Water Division
16 April

C.B. Whistfall Esq.
Hon. Secretary
West Central Association, ICE

Dear Mr Whistfall,

Your letter of 15 April arrived on my desk like a premonition of doom - I began to wonder whether your first initial stood for Cassandra!

However, despite having the unhappy feeling that I was approached as the last resort, I should be pleased to assist you in your dilemma and you may insert my name on the toast list. I should be grateful if you would let me have a note of the principal guests and their backgrounds. With regard to the refund of my dinner ticket, I would suggest that you send it to the Institution's Benevolent Fund as a donation in my name.

Yours sincerely,
R. Fitzjohn

Willpipe University
Civil Engineering Dept
22 April

R. Fitzjohn Esq.
Willpipe Water Division

Dear Mr Fitzjohn,

On the instructions of the Chairman I write firstly to thank you for your readiness to help last Friday and secondly to express his extreme disapproval for the manner in which you carried out your duty. He states that your speech took exactly 1 minute and 3 seconds to deliver; you mentioned only one guest by name, and that was the responder to your toast, and the rest of the period you spent upon an ironical apology to all present. Such antics can only bring disrepute upon the Institution for the important guests could well feel that they had been insulted.

Yours sincerely,
C.B. Whistfall,
Hon. Secretary

C.B. Whistfall
Hon. Secretary
West Central Association ICE

Dear Mr Whistfall,

Your letter of 22 April cannot be allowed to pass without reply. I am surprised both to hear that the Chairman's watch was working (he must have wound it up after he sat down) and also that I was speaking for over a minute - I must have been rather verbose.

I draw your attention to the fact that Dr Barnun, who ostensibly was present to propose the Toast of the Institution, instead took the opportunity to ride his hobby-horse upon the latest management gimmicks and jargon, discarded by the United States but unhappily adopted in this country with enthusiasm by some of the academic practitioners of the art. He also poured scorn upon civil engineers as managers in multi-functional bodies and spoke for 35 minutes by my watch (which was running all the time) apparently addressing his remarks to the members of the press who, until then, had been enjoying their free dinners at the corner of the top table.

The Chairman was obviously angry when he rose to reply, but he too compounded the grievance by speaking for another 31 minutes in refuting all the arguments put forward by the main speaker. The audience might well have enjoyed this treat, for our chairman has a mordant and sardonic wit when aroused, but unfortunately he chose to ignore the microphone and made his impassioned comments in a sibilant whisper to the culprit sitting beside him.

The other members and guests were becoming irritably restive and the hubbub from the back of the hall (rising with the inexorable speed of boiling milk) was seemingly ignored by the main protagonists.

As you had provided me with a list of 45 important guests (no wonder the tickets are so expensive) I quickly realised that I could not do justice to them all. I felt it would be prudent and subdolous to create a world record for the Guests speech in order to prevent a general exodus - were you not surprised to discover so many weak bladders among those present? My action may have been slightly reprehensible but, as you say, I did apologise. I was thanked by my responder later because it enabled him to give free rein to his fertile imagination and provide some recompense to the stoics who were keeping their patience under control with extreme difficulty. I might say that I have since received 67 unsolicited letters of congratulation.

Please note that, at the next Annual General Meeting of the Association, I shall be proposing a formal motion to the effect that we break with tradition in the future and invite ladies to accompany members to the next Annual Dinner. This should prevent a recurrence of last Friday's fiasco and should prove popular with the younger members.

Yours sincerely,
R. Fitzjohn

A crowd of men around our van, shaking their fists and beating the doors. What happened?

16

INTERNAL MEMORANDUM
3 January

To: Operations Manager   From: Divisional Manager

The telephones all seem to be dead today and I have not been able to get in touch with you. On my way to the office I went through the City Centre to see what progress was being made on the burst on the 450 mm high pressure main at the Castle cross-roads. I could not stop, owing to traffic congestion, but I could see a crowd of men around one our vans, shaking their fists and beating the doors. What happened down there this morning?

INTERNAL MEMORANDUM
6 January

To: Divisional Manager   From: Operations Manager

With reference to your enquiry about the Castle crossroads burst, I do not know whether to laugh or curse. The Jointer in charge was Harry Toothill and he was in difficulties trying to fit a collar on the make-up piece. There was a cable running across the joint and he said that from its' size he knew it was one of the disused tramway cables that were never lifted. His mate tried to dissuade him but Harry said he knew best and got to work with a hacksaw. There were a few sparks and, as far as he was concerned, this proved that he was right. He told his mate that if it had been an electricity cable he would have been dead but those old tramway cables all had a bit of electricity locked away in them and it was just draining out when he cut it.

As you have guessed, it was a main telephone cable going into the exchange and 2000 subscribers (including us) are now cut off. When the GPO men found out what had happened they tried to get at Harry but he locked himself in his van and

refused to come out for six hours. The GPO say that they will have to lay 2000 ft of new cable between manholes and I shudder at the bill to be met by our Insurance Company.

Toothill's nerve has now snapped and he has applied for a light job. There is a vacancy for a yardman at the main depot and I suggest he be transferred.

INTERNAL MEMORANDUM
7 January
To: Operations Manager   From: Divisional Manager

I agree to your proposal that Toothill be transferred to the main depot as yardman. He is over 60 and has worked in the industry for 45 years but I know that he does not relish the thought of retirement. The yard is in a bit of a mess - get him to clear it up and organize it properly.

INTERNAL MEMORANDUM
16 January
To: Operations Manager   From: Divisional Manager

The telephones were reconnected this morning and worked for an hour then they went dead again and I still cannot get in touch with you. When I left the office en route for lunch, I passed the depot gates and saw a familiar sight - two men hammering on the doors of a van. What is the story this time?

INTERNAL MEMORANDUM
17 January
To: Divisional Manager   *From: Operations Manager*

Re your memo of 16 January, I feel you know what I am going to say. Harry Toothill, who jumps at the sound of his shadow hitting the footpath these days, was trying to keep the depot tidy and, although he is not familiar with the the controls, tried to move the yard crane into its niche with the jib vertical. He smashed into all the telephone wires, brought them down and was so shaken that he slewed round and knocked a drum of waste oil over the road.

The Superintendent sent him off in a van to enlist the aid of the GPO and to purchase a fresh stock of tranquillizers. The Post Office engineers arrived in a car in about half an hour, closely followed by Toothill. The GPO driver tried to brake on seeing the oil, slid forward and bashed the front of his car on the brick wall. Ten seconds later Harry also skidded on the oil and hit the back of the GPO car just as the men were getting out. He then locked himself in his van for a further six hours and seemed to be chewing all the time. The telephones will not be repaired until Monday and Toothill says that he is applying for a job at HQ as handyman in the reprography department. (He may know nothing about this work but he is certainly handy as he lives around the corner.)

INTERNAL MEMORANDUM
17 January
To: Administration Officer   From: Divisional Manager

I note that HQ have sent us no Telex messages and no question-aires for the past week. Did Toothill get the job in the reprography department?

The two star-crossed lovers fell into each others' arms

17

To: Divisional Manager   From: Administration Officer

Welcome back after your holiday. We have kept a relatively clear desk for you but I attach three letters that have put me into a quandary and I thought you would prefer to deal with them!

Letter 1

Dear Sir,

I visit Willpipe about once a month to see my widowed mother and to carry out any odd jobs in the house or garden that she cannot manage herself. Also to settle any problems that have arisen with her next door neighbour who is a misanthropic bully. On Saturday I was burning some garden rubbish when this oaf appeared at the hedge with a hosepipe and not only extinguished the fire but soaked me into the bargain.

As I know that the Willpipe Water Division is operating a ban on the use of garden hosepipes, I wish to report him for contravening the ban and will happily appear as a prosecution witness. His name is Clarence Higbold and he lives at 17, Buckingham Gardens, Willpipe.

Yours faithfully,
Edmund Petty

Letter 2

Dear Sir,

I wish to complain about one of your water inspectors, J.D. Colclough, who tried to gain access to my house last Friday at 10.30 pm on the pretence of examining the plumbing installation. I am a single woman and there have been cases of rape in this area recently. We should be protected against people like Jim Colclough and the water authority should know what is going on.

Yours faithfully,
Roxanne Hart (Miss)

Letter 3

West Central Water Authority

To: Manager, Willpipe Division   *From: Director of*
                                            Scientific Services

The Authority has decided to co-operate with the Institute of Medico-Technical Research in determining whether water supply has any influence upon the birth of twins. Please complete the attached questionnaire listing the numbers of families with twins for each source of supply together with chemical analyses of the water supplied.

You will note that questions 4, 5 and 6 deal with the religion of the parents, the per capita water consumption for each house and the type of material used in the service pipe and obviously each house will need to be visited by a responsible official. The data should be returned to me within 15 days.

INTERNAL MEMORANDUM
15 July

To: Administration Officer   From: Divisional Manager

The three letters attached to your note of the 14 July indicate that the silly season has commenced. I return them herewith and they should be answered as follows:

Letter 1 Please inform Mr Petty that, although we are operating a hosepipe ban, we still have a statutory duty to provide water freely for fire-fighting purposes and therefore any prosecution pursued against Mr Higbold would inevitably be doomed to failure.

Letter 2 Something stinks with this letter despite Colclough's well known amorous proclivities. Please pass the complaint to the Distribution Engineer who should arrange to call on Miss Hart by appointment accompanied by Colclough and by the inspector's Trade Union representative. I shall want a report on the outcome of the meeting.

Letter 3 This is an obvious hoax which was carried out in the absence of Mr Staunch, the Director of Scientific Services, who has been on holiday in Ibiza. I would agree, however, that it is no no more absurd than some of the other questionnaires received during the past 12 months. We must make the hoax backfire and I suggest you reply to the Director next week on the following lines:

Deepwet Reservoir Source

Number of domestic consumers	50 310
Number of family units	15 617
Number of families with twins	79
Religion of parents:	33 Church of England
	23 Roman Catholic
	2 Mohammedan
	1 Buddhist
	8 Noncomformist
	10 Agnostic
	2 Jewish
Material of service pipes	61 copper
	4 lead
	3 galvanized iron
	11 polythene
Additional temporary staff appointed	1 No. Grade 3 supervisor
	2 No. Grade 1 clerks
Cost of survey to date	£2300
Future cost of survey (all sources)	£16 500
Estimated cost of meter installations	£1580

Please remit approval of overall expenditure in sum of £20 380. There should be a firework display at Headquarters when Mr Staunch gets your reply. Amend any figures that seem unrealistic as I put down the first ones that came into my head.

INTERNAL MEMORANDUM
21 July

To: Divisional Manager   From: Distribution Manager
Following your instruction, I arranged an appointment with Miss Hart for 11.30 am on Saturday morning together with Jim Colclough and Bill Brewer (his shop steward). I had first interviewed Colclough in the presence of Brewer and he was

amazed at the complaint. Apparently, to use an old-fashioned term, he and Miss Hart had been courting for the past 12 months although he openly admitted that the courtship was all on her side and he was just interested in the fringe benefits which consisted of sleeping with her on Fridays and Saturdays. On the previous Sunday they had quarrelled because she was getting annoyed at his attitude. He had not seen her during the week but called at the house late on Friday night to make amends. An elderly man, a stranger, opened the door, asked him who he was and what did he want. He was so taken aback (and he had had a few drinks that night) that he stammered that he was a plumbing inspector from the water division whereupon the door was slammed in his face.

The Saturday interview was a most peculiar experience because I had previously informed Miss Hart that her complaint was being investigated and there would be serious consequences for Colclough if her allegations were substantiated. As soon as the door opened, these two star-crossed lovers (to use another cliché) fell into each others arms while Brewer and I felt completely de trop. We managed to break them up before the neighbours complained and all went into the house for a glass of sherry and an explanation. Apparently Miss Hart had not expected Colclough to call on the previous Friday and was entertaining her aunt and uncle. The uncle, a man of a choleric disposition, answered the door and, receiving Colclough's limp explanation, later irascibly insisted that Miss Hart write you a letter of complaint. She being unwilling to admit her liaison with Colclough and still being annoyed with him, complied and her uncle posted the letter on his way home. I agreed that the letter would be destroyed and Brewer and I departed as quickly as we could, Brewer remarking that he wished all management and union problems could be solved so easily. Somehow I do not think much sherry was drunk after our departure and despite Colclough's views, I expect you will have to contribute to another wedding present list in the near future.

18

To: Divisional Manager   From: Administrative Officer

You will recall that some time ago, after receiving several complaints from me regarding the inefficiency of the new Administrative Assistant, Mr Pinchbeck, you interviewed him, gave him a severe reprimand and put him on probation for three months. This period has now elapsed and, as instructed, I give below my report on his progress.

I must admit that I was originally rather dismayed by your decision and Mr Pinchbeck's work during the first two weeks reinforced my views. However I then discovered the BL principle and have since given it a trial run during the remaining ten weeks with marked success. You may have noticed that my department has worked smoothly and efficiently in the last couple of months and I have not bothered you with any tiresome problems?

Mr Pinchbeck remains idle and inefficient; he possesses the brain power of a mouse and his lack of judgement strains credulity but nevertheless I recommend that he be promoted one grade and appointed to a new post in the Establishment, designated Special Assistant (BL). I also request permission to advertise for his successor as I still require another administrative assistant.

Before you reach for your telephone to find out whether I am mad or impertinent or both, I had better explain the BL principle to you in detail.

You may or may not have known, (but I can assure you that it is correct) that the only way a gambler can consistently win at roulette is by finding which other player at the table is in desperate need of money and then betting against him. For example, if the desperate player backs red then the successful gambler puts his money on black. If the player selects odd,

Whenever a problem arises I take it to Mr Pinchbeck, ask for his decision and do the opposite

the gambler selects even and so on. In other words, the desperate player is a Born Loser. This principle applies in other spheres of life - I feel sure that you can think of one notable example in politics. Born Losers are scattered throughout the population and they must be there for a reason. It is said that a successful manager is one who makes the right decision more than 65% of the time while a genius has a success rate of 75%. By applying what the statisticians call a reverse correlation to the decisions of a Born Loser, I firmly believe that we can achieve hitherto unknown success rates of over 90%.

At any rate, the system works with Mr Pinchbeck. We have put him in a room by himself and supply him with tea, cigarettes and the daily newspaper. He spends his time trying to do the crossword puzzle and has yet to complete the simplest one. Whenever a problem arises, I take it to Mr Pinchbeck and ask for his decision. I get his answer and then do exactly the opposite. So far he has not failed me and I realise that in Pinchbeck we have probably recruited one of the prime BL Assistants. I have no doubt that he is destined for higher things and once Headquarters discover the BL principle and adopt it, I can see him being moved from the Divisional Office. However recruitment of BL Assistants is relatively easy and foolproof, although perhaps time-consuming. One merely asks each candidate to call 'heads' or 'tails' when tossing a coin 100 times in the air. Born Losers can then be graded as follows:

50 incorrect calls:	Average
60 incorrect calls:	Latent BL tendencies
70 incorrect calls:	BL Assistant (BLAST)
80 incorrect calls:	Senior BL Assistant (SEBLAST)
90 incorrect calls:	Prime BL Assistant (PRIMBLAST)

I have a suspicion that a score of 100 would signify the end of the world. To corroborate the foolproof nature of the above method of selection I would draw your attention to various MCC captains in the Australian cricket matches. They were obviously BLASTS or had latent BL tendencies. To guarantee continued success, it would be prudent to insist that BL assistants re-take the above test at weekly intervals to ensure that they are still mis-firing on all cylinders. For your information, Mr Pinchbeck checks out regularly with a score of 91.

You will appreciate that a BLAST is more effective when he is unaware of his potentialities and it follows that PRIM-BLASTS are near-morons. I should imagine that ambitious and intelligent BL assistants could be quite dangerous if they reached positions of responsibility and insisted on taking their own decisions. They might be termed DAMBLASTS and obviously, from the state of the world, there are a lot of them about.

If this principle is followed up on a national basis, the United Kingdom could be put on its feet once again by utilizing the talents of specialist BL assistants in the Diplomatic Service (DIMBLASTS) and the Treasury (TREBLASTS) while as officers engaged on astronautical research they would be known as BLASTOFFS.

There is no doubt at all in my mind that the world is due to enter upon a new era of prosperity when the BL principle is fully developed. My one sorrow is that the principle cannot be patented. I will leave it to you to start the ball rolling.

19

To Director of Operations   From: Manager
West Central Water Authority   Willpipe Water Division
As you know, whenever I attend a group meeting on your behalf,
I provide a brief summary of the proceedings for your information. I regret to say that I must give a 'Nil' return for the
progress achieved last week. The reason generally for this result
was that, by some mischance, the members present (apart from
the Secretary and myself) were comprised wholly of eccentrics
and professional committee men. To amuse myself on the return
train journey, I wrote thumbnail sketches of the various types
who attended and you might like to try and put names to the
portraits. They were:-

The Earnest Bird.   He loses sight of the main principle at issue
but will defend an oblique view of a minor item to the death. He
generally raises the trivial point late in the meeting and at his
best (or worst) can nullify the discussion of the previous 3 hours.
Also keen on raising 'Points of Order'.

The Euro-Committee Bird.   Fond of quoting EEC directives
and European resolutions to prove that the group cannot achieve
its objective, he can reduce his colleagues to abject despondency
as they begin to realise that their counter-arguments are
encountering a cushion-like indifference. Usually some
members are lulled to sleep by the confident drone of his voice
and it is these fortunate beings who, having closed their minds
and remained unaffected, bring the meeting back to order when it
appears to be drifting into obscurity.

The Musical Gourmand.   No one knows why he is a member
of the Committee because he neither vouchsafes an opinion nor
votes on an issue. Sometimes it is difficult to discover the
organization he represents. He usually arrives late without his
papers and scrounges spare copies from the Secretary. He eats
and drinks twice as much as anyone else and invariably leaves to

THE EARNEST BIRD

Point of Order

THE EURO COMMITTEE BIRD

AGENDA

THE LOBBYIST

THE HOBBY HORSE

DIARY

THE PERPETUAL MEETING MAN

THE PSUEDO ILLITERATE

THE MUSICAL GOURMAND

Various types who attended

76

attend a concert of chamber music. However he has one good facet—he may not contribute but neither does he hinder.

The Hobby Horse. Generally a research scientist who manages to drag a particular bee (from his bonnet) into every discussion. Invariably sits near a blackboard so that he may illustrate his remarks in chalk. At an institution meeting he can be identified because he will be carrying a bulging brief-case and a box of slides. Always possesses 30 copies of each piece of paper for distribution to his colleagues.

The Lobbyist. He has no interest in 95% of the matters under discussion but is determined to obtain a decision on one point only. This point is generally located at the end of a lengthy agenda and, because he fears that it may be deferred, he will entreat the chairman to change the agenda with the plea that he needs to catch an early train for another important appointment.

The Pseudo-Illiterate. He either has not or could not read a single word of the introductory papers and reports and, arriving with a blank mind, disguises the fact by asking the chairman to take the papers paragraph by paragraph to enable him to assimilate the interpretations. This cuts the effective discussion by 50% and can reduce it to a dialogue.

The Perpetual Meeting Bird. Sitting as he does on some 30 or 40 committees considering overlapping subjects, he has a full diary and finds difficulty in agreeing a date for the next meeting. He enjoys referring items to his other committees, thus delaying progress and ensuring full rotating agendae for his many meetings with the added advantage of being able to pose as an expert on the subject when it is next raised.

The After-thought Bird. Special meetings are called at the request of this person who gets a change of heart two days after a closely-fought issue has been decided, writes to the Secretary with a long carefully-reasoned diatribe and asks that the matter be reconsidered in view of the fresh evidence. The Secretary can never discover the fresh evidence but the bird has plenty of supporters to reinforce his views.

The Establishment Observer. A sardonic term to describe an inveterate speaker who rejoices in specifying the obstacles and difficulties, carries no responsibility and can suggest no solutions. It is almost apodictic that he was deliberately nominated to introduce sand into the wheels of progress.

The Ineffectual Chairman. One law attributed to Parkinson states that a person gets continually promoted until he reaches (or over-reaches) a level where he is incompetent. It is incontrovertible that such a person will eventually sit on an excessive number of committees and will undoubtedly be chairman of several by reason of his seniority. He is fond of saying 'I'm in the hands of the committee' to disguise his lack of opinion and often argues at cross-purposes with his members. In accordance with Sod's Law he only takes a firm stand when he is utterly wrong. If the Secretary makes the mistake of allowing him to see and amend the draft minutes, he is capable of setting a committee back three months.

No doubt Offa Dyke could think of many more such types and write much better descriptions of them. I invite him to continue. In the meantime, the presence of these characters does much to explain why the forewords to Government Reports often state that the committee met on 57 occasions during the past four years and also why the lists of committee members contain so many resignations and fresh appointments.

20

Deepwet Angling Society

Engineer and Manager
Willpipe Water Board
Dear Sir,

The Deepwet Angling Society have instructed me to draw your attention to the behaviour of the hooligans sailing on the reservoir. Since you last intervened two years ago the situation has steadily worsened. There is so much skylarking by youngsters without adequate supervision that the engine of the rescue boat is never silent as it zooms up and down the reservoir attending to the dinghies that have capsized. The peace of the countryside has been ruined and the fly fishermen find it impossible to enjoy any recreation as the boats sail so close to the banks.

I have informed the Sailing Club that we can no longer tolerate such a nuisance and I must request that the Water Board gives the Sailing Club notice to quit.

Yours faithfully,
K.F. Jones
Hon. Secretary

Deepwet Sailing Club

Engineer and Manager
Willpipe Water Board

Dear Sir,

I regret having to complain once again about the fishermen using Deepwet Reservoir. They appear to think that they alone have a right to use the reservoir for recreational purposes. They intimidate the Junior section of the Club if no senior members are in the vicinity, they puncture and sink the

The situation has steadily worsened

marker buoys and boundary floats and their abusive language has shocked some of the lady members.

I have informed the Angling Society that it is unthinkable that a few fishermen should be allowed to ruin the enjoyment of so many yachtsmen and the Sailing Club has instructed me to request the Board to revoke all fishing licences and thus save money by not restocking the reservoir. Please let me know when the fishing will be terminated.

Yours faithfully,
L.M. Centreboard
Hon. Secretary

Willpipe Water Board

L.M. Centreboard Esq
Hon. Secretary
Deepwet Sailing Club

Dear Sir,
    I was sorry to receive your letter demanding that the Board terminates fishing on Deepwet Reservoir. I knew that the two organizations were in dispute but I thought that two such reasonable men as you and Mr K.F. Jones would be able to work out a solution especially as I know that you personally hold such a high opinion of Mr Jones.
    As it is, if one recreation has to stop I agree it will have to be the fishing as the Sailing Club have enjoyed good relations with the Board for such a long time.

Yours faithfully,
Engineer and Manager

Willpipe Water Board

K.F. Jones Esq
Hon. Secretary
Deepwet Angling Society

Dear Sir,
    I was surprised to hear of the disputes arising between your

members and those of the Sailing Club especially as you always speak so admiringly of the organizing ability and tolerance of Mr Centreboard.

However, if only one organization can remain on the reservoir I agree that the sailing will have to cease as the Angling Society have enjoyed good relations with the Board over such a long period.

Yours faithfully,
Engineer and Manager

File note by Engineer's Secretary
On the instructions of the Engineer and Manager, the previous two letters were deliberately posted in the wrong envelopes.

Deepwet Angling Society

Engineer and Manager
Willpipe Water Board

Dear Sir,

I am pleased to inform you that I happened to meet .Mr Centreboard, the Secretary of the Sailing Club, recently and following an interesting discussion with him we have managed to sort all the differences between the Sailing Club and the Angling Society. I should be grateful therefore if you would disregard all previous correspondence on this subject.

Incidentally I return herewith a letter (unread) that you have obviously sent to the wrong address.

Yours faithfuly,
K.F. Jones
Hon. Secretary

Deepwet Sailing Club

Engineer and Manager
Willpipe Water Board

Dear Sir,
My secretary tells me that we received a wrongly

addressed letter at this office and I have asked her to return it to you in this enclosure.

You will, I am sure, be interested to know that I bumped into Mr K.F. Jones yesterday and quickly managed to iron the trivial disputes that had been bothering our respective members. I would therefore ask you to take no action on any previous correspondence on this subject.

Yours faithfully,
L.M. Centreboard
Hon. Secretary

Communication reticulation

21

To: Manager   *From: Director of Operations*
Willpipe Water Division   *West Central Water Authority*

*In confirmation of our telephone conversation today, I find that
I shall be unable to attend a three day seminar in London on
'The Philosophy of Management Training' and you have
kindly agreed to take my place. I enclose all the relevant
documents and I should be grateful if you would let me have a
brief report on the seminar in accordance with Authority
Procedure when you return to Willpipe.*

To: Director of Operations   From: Manager
West Central Water Authority   Willpipe Water Division

I returned to Willpipe yesterday, following the seminar, feeling
that my head was filled with boiled cotton-wool, but I have now
recovered sufficiently to let you have my thoughts while they are
still fresh.

I travelled to London with an open and perhaps naive mind,
my only reservation being the improbability of anyone being
able to discuss 'The Philosophy of Management Training' for
three days. In my simple way I had considered a good manager
to be a natural leader with excellent judgement and a fund of
sound common sense, while a definition of training could be
reduced to three words: Demonstrate, Teach, Supervise.

However, at the seminar there were six main speakers (all
seemed to have the academic approach and be under the age of
30) and 17 'outsiders' (none of whom had less than 20 years
experience of management). Without exception, my 16 colleagues did not understand the dissertations and the only one
who enjoyed himself was a company commander from a
regiment stationed in Ulster who had travelled over for a rest and

change of scenery, and regarded the whole exercise with amused tolerance.

My head started to whirl and my heart sank to my boots when the opening speaker stated that 'Consummation of time in discussion is validated by the gain in commitment of the participants'. I fear that my commitment has regressed instead. I have noted some of the jargon spouted by different lecturers and I can but apologise if I have made some mistakes in writing it down as my brain was beginning to regard my ears with suspicion. Nevertheless I feel that errors could only improve nonsense such as this:

(a)   The subjective spectrum may be eidetically retained by ingesting a wider awareness of the prescription of training needs analyses.

(b)   The objective appraisal of facilities can be derived in special workshop seminars from a critique and a quasi-correlation of the functional areas of management, the behavioural sciences and the quantitative aspects of decision making.

(c)   Foundation of the methodology is a pre-requisite to the systematic strategies of executive programming.

(d)   The harmonization of process and flexibility is a polymorphous concept postulated as the nexus of managerial efficiency.

The speakers also had a predilection for enclosing words in boxes, drawing lines between them and describing the resulant maze as a 'communication reticulation'.

I cannot understand the above phrases and would welcome some lucid interpretations but, if this seminar is typical of the rash of discussions, courses and study-groups that is erupting over the whole industrial scene, I am beginning to understand why this country finds itself in its present parlous plight. No doubt Offa Dyke would agree with me. I naturally complained at the conclusion of the seminar but the chairman condescendingly informed me that I could not be expected to grasp the subject in 3 days. He intimated that 3 weeks would be preferable. Personally I do not consider that I possess sufficient resilence to endure 1 week, let alone 3.

After a gap of 24 hours and a return to the real world of

personnel problems and Sod's Law, I am quite unrepentant. I believe the seminar to be a waste of time and money and wish to be excused from any similar chores in future. I firmly believe that this mass of verbiage is cloaking the emergence of a new profession whose sole productive influence consists of teaching others to join it and I suggest, in conclusion that a mistake was made in the name of the seminar (the speakers would no doubt call it a 'Freudian transposition'), the true title being 'The Training of Management Philosophers'.

This machine has been retired from service and I have given instructions for it to be scrapped

22

Halcyan Excavators Ltd
Barchester
13 April 1979

Engineer and Manager
Willpipe Water Board

Dear Sir,

      Our records indicate that your Board purchased a Model 321 Halcyan excavator from us in 1975. In connection with the sales campaign we are launching in June, it is proposed to issue a publicity pamphlet on the Model 321 giving a technical specification and including comments of satisfied users.

      As one of our oldest customers, I should be grateful if you would let me have your observations for inclusion in the booklet.

Yours faithfully,
S.S. Hardcell
Managing Director.

Willpipe Water Board
15 April 1979

The Managing Director
Halcyan Excavators Ltd
Barchester

Dear Sir,

      With reference to your letter of 13 April, if you wish to publish my observations on the Halcyan 321 Excavator you are welcome to them.

      This machine was a complete disaster from the day it was delivered. It was too light, underpowered, unreliable and must have been designed by a Chinese Heath Robinson. The digging chains and bucket teeth broke on the second day it was used and we were unable to obtain spares for three months. Follow-

89

ing repair, it limped along for another three days before the boom cracked. It took four months to obtain another boom and then we obtained a week's work out of it in sandy ground before the engine packed up. By this time, according to the small print on your quotation, we were out of the warranty period and we decided to cut our losses. Accordingly it has been unused since February 1976. I have consistently warned other potential purchasers of the propensities of this machine.

> *Yours faithfully,*
> *Engineer and Manager*

Halcyan Excavators Ltd
Engineer and Manager   19 April 1979
Willpipe Water Board

Dear Sir,

I was sorry to hear of your experiences with the Halcyan 321 Excavator. You had obviously purchased one of the original prototype machines in 1975. Since that date many modifications and improvements have been made to the basic design and the Company itself is now operating under new ownership and new management.

To retain your goodwill, I have arranged for our new mobile workshop to call at your depot next Thursday to modify and restore the machine to full working order at no cost to yourself.

> Yours faithfully,
> S.S. Hardcell,
> Managing Director.

Halcyan Excavators Ltd
Barchester
Engineer and Manager   29 April 1979
Willpipe Water Board

Dear Sir,

I have now heard from my Maintenance Manager that

your Halcyan excavator was completely repaired by Tuesday, 27 April. No charge will be made for this service but you might be interested to know that it cost this company £455.

I trust that I can now look forward to some, more favourable, comments from you.

Yours faithfully,
S.S. Hardcell,
Managing Director

The Managing Director
Halcyan Excavators Ltd
Barchester

Willpipe Water Board
10 May 1979

Dear Sir,

With reference to your letter of 29 April, I confirm that the Halcyan 321 was repaired on 27 April. It was put to work on the Wednesday and by Friday the digging chains and bucket teeth had broken. These were replaced (I found that your mobile workshop had sold certain spares to my plant foreman) and the excavator was started again on Monday, 3 May. On Thursday the boom collapsed and my patience evaporated. The machine has been retired from service and I have given instructions for it to be scrapped.

In my opinion, the excavator will need to be completely redesigned before it can reliably carry out all its specified tasks and out-perform other reputable machines.

Yours faithfully,
Engineer and Manager.

INTERNAL MEMORANDUM
23 June 1979

To: Clerk and Solicitor   From: Engineer & Manager

I attach copies of correspondence with Halcyan Excavators Ltd., together with a copy of their latest brochure. Is there nothing we can do about this sort of thing?

You will note that, despite all my condemnations, they state in the brochure:

"The Model 321 Halcyan excavator purchased in 1975 by the Willpipe Water Board has received only one overhaul in 4 years. The Engineer to the Board states 'it can reliably carry out all its specified tasks and out-perform other reputable machines".'